Counting Concepts Count

Digging Deeper
Can You Dig It!

Created by Daryl Kuiper

Copyright 2014 by Daryl Kuiper

Contact the author:
darylkuiper@gmail.com
mathreaders.blogspot.com

ISBN-13:978-1507805565
ISBN-10:150780556X

Special thanks to Kenneth Guentert and
Lynnee Sullivan Kuiper

This book belongs to

_____

Howdy, It's me again, Seven.
For those who want a deeper understanding of Counting Numbers, this is your book.

We will talk about number games and the words represent, representation and concept.

Wow!
This will be fun.

There are lots and lots of us Counting Numbers.
Did you know there is no biggest counting number!
Think of the biggest number that you can.
What is it?

Now, I bet you can imagine the next number after that, just one step bigger, especially now that you think about it.

You are playing the Next Number game.
You can play that game with your
family or friends.
If I say the number **one**, you would say
_____.   then I would say **three**
and you would say _____.

I started out
simply because I wanted
you to get
the idea.
Can you continue?

You can start anywhere. You could start
with **six**
(if you wanted to) or me, **seven,**
or One hundred twenty five, if you know
that big number already.

None of us Counting Numbers can claim to be the greatest or biggest number. Why? Because there is always another Counting Number to the right that is bigger. The Next Number game will never end. That's right, you can count us Counting Numbers forever and forever and never be done.

The Next Number game.
It's a fun way to learn how to count really big numbers.

Just stop the Next Number game when you get tired.
You can never get a biggest one,
no matter how hard you try.
But go ahead and try if you want.

Represent

I use that word, represent, often.
What does that word mean?
Represent means to be able to draw or write symbols that help you think of something.... like us Counting Numbers...or even you!
When I say I am Seven, I mean I represent any set of seven things.

There are seven things in each of these boxes.

Is a picture of you really you?
A picture of you just reminds people of you, right?
A picture is just a representation of you.

In the same way the symbols used for us Counting Numbers are representations of us and not really us.

Picture of you

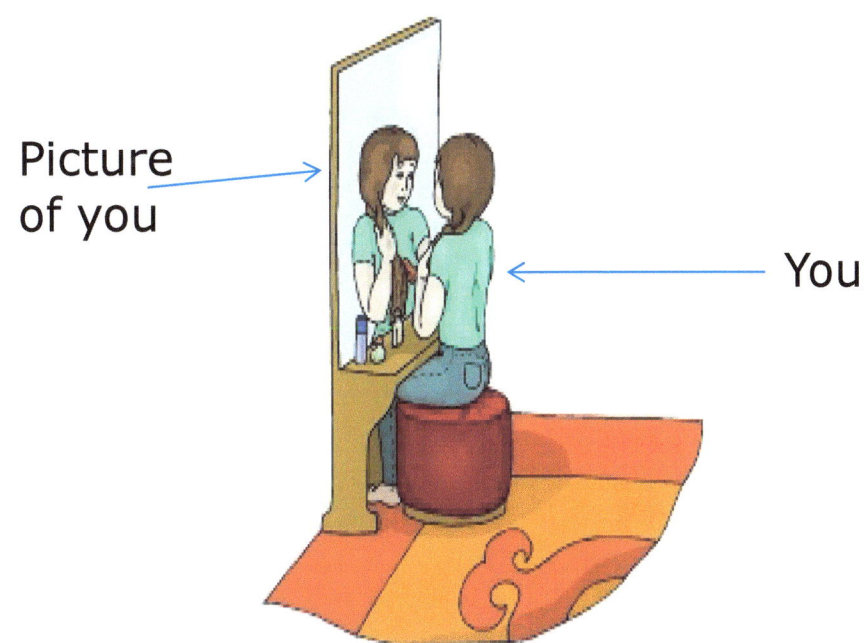

You

Remember how we could be drawn differently
in the other books or on the next page?

Remember One? Remember how he could
be drawn in different ways?

Don't forget his curviness Six and how he
could look.
These are all different representations
of us Counting Numbers.

That is what represent is all about.

Represent

Symbols

Those different ways of writing us Counting Numbers are called *symbols*. *Symbols* are used to represent us Counting Numbers.
Each of us Counting Numbers have a different name and we each have a different symbol <u>representation.</u>

1
6
7

Sometimes those symbols can be a little different.
Get it?
<u>Represent</u>.
Got it?

Symbols not cymbals!

Now here is the trick. You know we use words to represent us. Right? Well, in other languages, the words used to represent Counting Numbers are different.

Examples

In Germany, One is called
Ein,
In Spain, Two is called
Dos,
In Japan, Three is called
San.

Many languages of the world have different words to represent us Counting Numbers.
It is interesting to study languages and geography.

Even though each of us Counting Numbers can have a different name in a different language, each of us is still the same concept.
It might take a while to understand that last sentence. Think about it.

So now for the next trick. We Counting Numbers are represented by more than words. You know that, right? We also are represented by symbols, like me right here.

7, 7, 7, 7

The trick is to use symbols that you and others can understand.

We have seen how numbers can be represented by slightly different symbols. Can you remember some of them?

It can be even trickier. In countries like Japan and China, really different symbols are used.

This is the symbol both Japanese and Chinese writers use for me, Seven. Looks like a "t" does it not? But the concept of 7 is still there. Yup, I know, it's a little complicated.

When a Japanese child looks at this symbol it is like you looking at me, 7.

It takes time getting friendly with us Counting Numbers.
There is so much to learn; how to say our names, how to draw our symbols. and how to count things with us.

I think learning is exciting . What do you think?

I think learning is fun. How about you?

Seven

One

SIX

It takes time to learn about your friends too, doesn't it? But you can make it fun. I know you will learn our names and symbols (at least the English ones) in no time with some practice. It's like learning the names of all your friends.

1

6

7

Alisha

Juan

Marina

Tisha

Shawna

There is another way to play the Next Number game. Instead of saying the name of the next number, this time draw its symbol.
If I draw 1 then you would draw __
and I would draw 3 and you would draw ___.
I am sure you get the idea.

When do you quit, you ask?

When you get tired or run out of paper, silly.
Remember, we Counting Numbers don't ever end. We keep counting and counting.

How about the Even Number Counting game?

2,4,6,8...... EVEN numbers

Have you tried counting by 2's? It is another fun game. I say 2 and you say 4. Then I say 6 and you say____ I'm sure you have it now. It can be played by drawing the symbols also.

When does the game end? You know.... When you want it to.

The rest of us, 1,3,5,7,... are called
ODD numbers.
Isn't that odd?

I don't think we are odd at all.
When we think of odd as being different
(or special!), we are OK with being called "odd".
We like being different (or special).

You can play the

Odd Number Counting game also.

What number would you start with?

Do you have to start with One?
When can you stop?

It is so much fun counting on us Counting Numbers because there are so many different games you can play with us.

How about counting by fives?

Or threes?

How about counting backwards from 100 by threes?

5, 10, 15, 20 ...

3, 6, 9, 12...

100, 97, 94, 91, 88...

You can make up all
sorts of games.

Have fun.

Isn't it fun learning all these
new words and ideas?

It is fun for me to share
with you.

But, before I end this book, this is important: do you ever get called names? Sometimes it's OK. Sometimes it's funny. It is OK by us Odd Numbers to be called odd.

But being called a name you don't like can hurt.

Wimp

This is not OK.

What do you do if you get called names you don't like?

I hope you tell an adult. Talk about it with them.

Meanwhile, count and count some more.

www.ingramcontent.com/pod-product-compliance
Lightning Source LLC
Chambersburg PA
CBHW050432180526
45159CB00006B/2503